CLOSER LOOK

THE OZONE HOLE

Alex Edmonds

Franklin Watts
LONDON ● SYDNEY

© Aladdin Books Ltd 1996
Designed and produced by
Aladdin Books Ltd
28 Percy Street
London W1P 0LD

*First published in Great Britain
in 1996 by*
Franklin Watts
96 Leonard Street
London EC2A 4RH

A catalogue record for this book is available from the British Library.

ISBN: 0 7496 2447 7 (hb)
ISBN: 0 7496 3561 4 (pb)

Editor
Selina Wood
Designer
Gary Edgar-Hyde
Picture Research
Brooks Krikler Research
Front cover design
Karen Lieberman
Illustrators
Tessa Barwick
Gary Edgar-Hyde
Ron Hayward
Aziz Khan and Creative Land
James MacDonald
Alex Pang
Mike Saunders

Certain illustrations have appeared in
earlier books created by Aladdin Books.

Consultant
Graham Peacock

Graham Peacock is senior lecturer in science
education at Sheffield Hallam University. He has
taught in schools in England and abroad and has
written science books for children and teachers.

CONTENTS

INTRODUCTION

Since the early 1980s, the "hole" in the ozone layer has become an issue discussed by scientists and politicians alike. There is now conclusive proof that we are destroying the ozone layer with human-made chemicals. So the race is on to repair the damage already done, and prevent future environmental disasters.

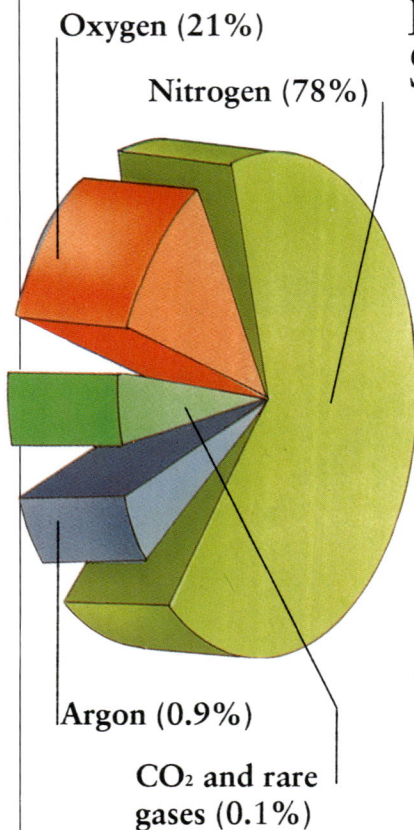

Oxygen (21%)

Nitrogen (78%)

Argon (0.9%)

CO_2 and rare gases (0.1%)

The Earth is thought to be the only planet in the solar system that supports life. Its atmosphere provides some of the conditions needed for life. It traps heat and it contains air, as well as a layer of the gas ozone, which protects us from the extremes of the Sun's harmful rays.

LIFE ON

ALL WRAPPED UP

The atmosphere around the Earth could be compared to a blanket. It contains certain gases which keep the Earth at the right temperature. Without these "greenhouse" gases, the planet would be too cold for all forms of life. Some of the heat that the Earth receives from the Sun is trapped by these gases as it is reflected back into the atmosphere. This works in much the same way that a greenhouse keeps its plants at the right temperature.

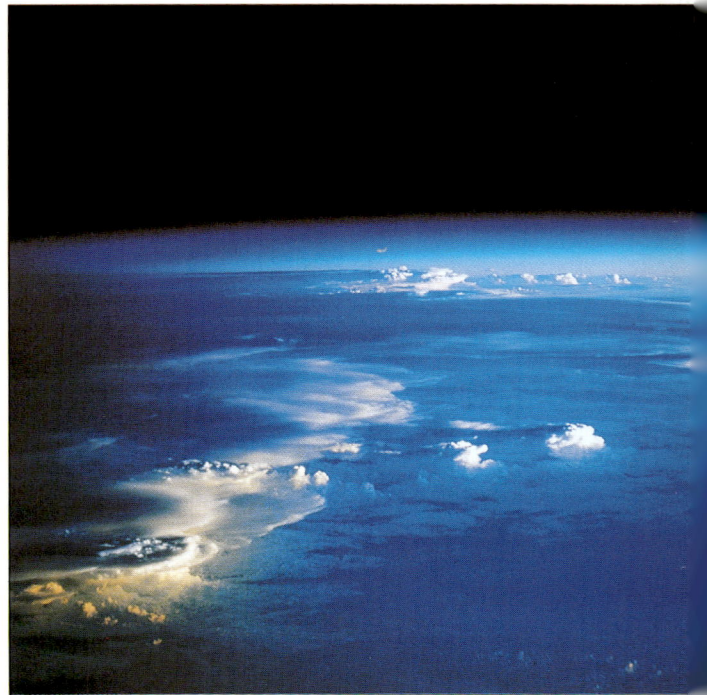

Rainclouds gathering over New Guinea. The photograph shows how the atmosphere resembles a blanket of gases over the Earth.

ON CLOSER INSPECTION
– Gasping on Everest

Most climbers on very high mountains, such as Mount Everest in Nepal (right), need breathing equipment. This is because the air at this altitude is much thinner and contains less oxygen.

EARTH

THE ATMOSPHERE

There are four main layers in the atmosphere. The troposphere, which extends for about 11 km up from Earth, contains breathable air. This is where hot air balloons and gliders travel. The next layer, the stratosphere, reaches 50 km above Earth. Here we find the ozone layer, and in the lower stratosphere aeroplanes cruise. The mesosphere and the thermosphere extend to 500 km. After that we find the exosphere, the outer limit of the atmosphere. Beyond that is space!

Satellite

Exosphere

Meteorites

500 km

Aurora Borealis

Space shuttle

Thermosphere

80 km

Ozone layer

Weather balloon

Mesosphere

50 km

Stratosphere

Air balloon

11 km

Glider

Troposphere

gamma rays

x-rays

ultraviolet

visible white light

infrared

microwaves

radiowaves

The Sun is the nearest star to Earth and is at the centre of the solar system. Life on Earth relies on energy radiated from the Sun, known as solar radiation. It is vital because it provides us with light and heat. One type of solar radiation is ultraviolet (or UV). It is this ultraviolet radiation that the ozone layer protects us from.

LIFE - GIVING

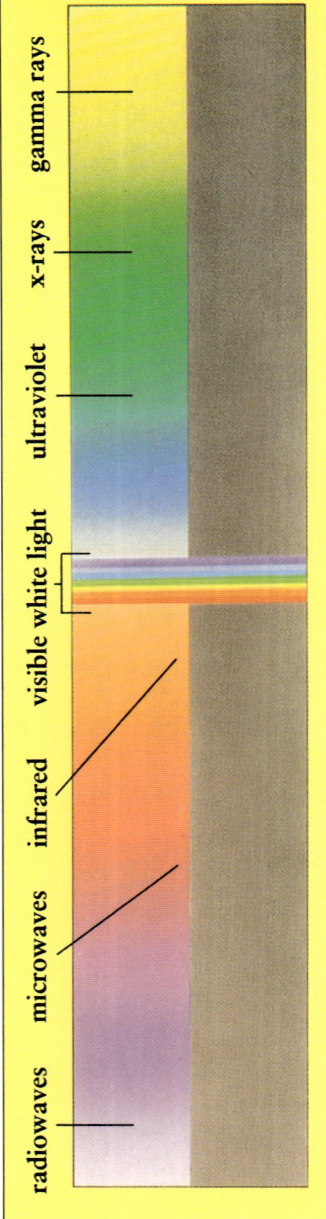

Electromagnetic spectrum
There are many types of radiation. We see white light. Gamma rays and x-rays are used for medical screening. Ultraviolet is used in sunbeds, infrared in remote control, microwaves for cooking and radiowaves for radios.

KEEPING US ALIVE
On a warm day we can feel the Sun's rays. Plants need sunlight (as long as it is not too intense), using the Sun's energy to grow (see page 19). Without plants there would be no life on Earth. The oxygen we breathe comes from plants. So does our food, either directly, or indirectly from animals who eat plants.

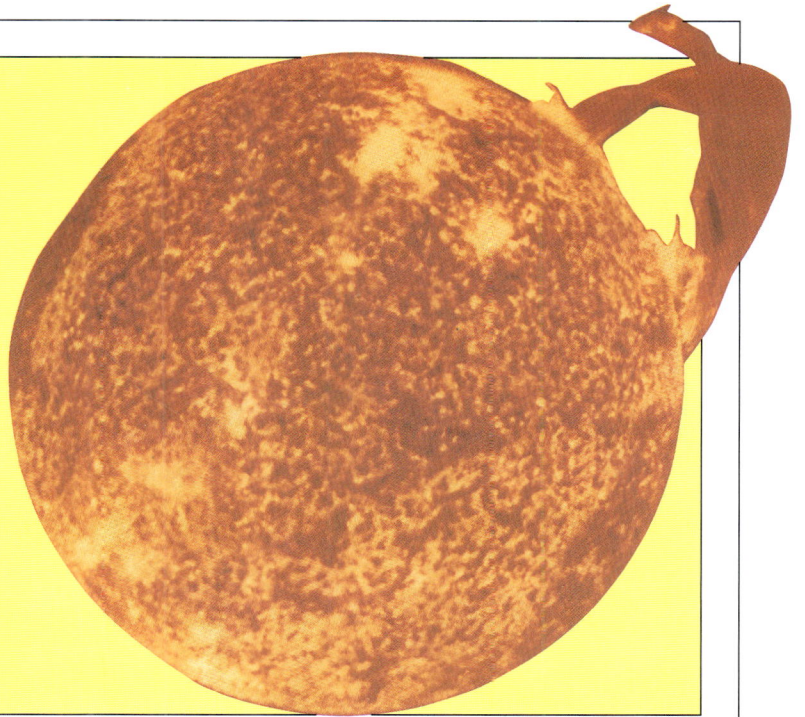

SUN

SOLAR RADIATION

The Sun's radiation travels the huge distance from the Sun to us. As it does so, much of the energy is lost. On reaching our atmosphere some energy is absorbed and some is reflected back. Closer to Earth, clouds also reflect and absorb energy. What remains hits the ground and is again reflected or absorbed. We therefore receive only a part of the radiation that travels from the Sun. Scientists are developing new ways to use more solar power – to generate electricity, to heat and cool buildings and to power cars. Solar power may become a major energy source in the future.

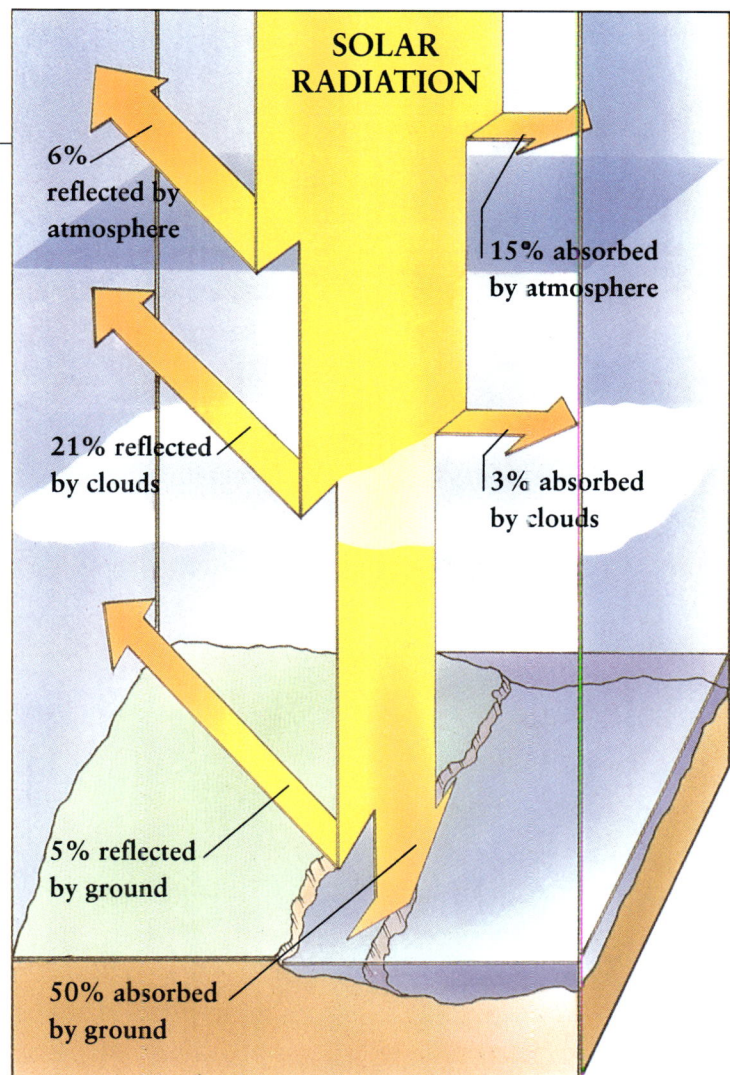

SOLAR RADIATION

6% reflected by atmosphere

15% absorbed by atmosphere

21% reflected by clouds

3% absorbed by clouds

5% reflected by ground

50% absorbed by ground

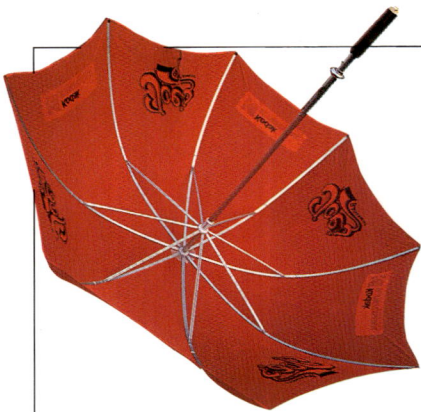

Sun-shield

A good way to think of the ozone layer's job is as a Sun-umbrella over the Earth which protects us from the Sun's powerful rays.

The ozone layer is in the stratosphere, between 11 and 50 km above sea level. Only a millionth of the stratosphere is the gas ozone, but it is essential. It filters out most UV-B radiation, which is a small part of ultraviolet, that causes sunburn.

THE OZONE

THE OZONE LAYER

When UV-B meets ozone in the atmosphere it is absorbed by the ozone, and at the same time breaks down the ozone into a different form of oxygen. These forms then unite with others that have also been broken down, and reform as ozone. Under normal circumstances the amount of ozone in the ozone layer remains constant.

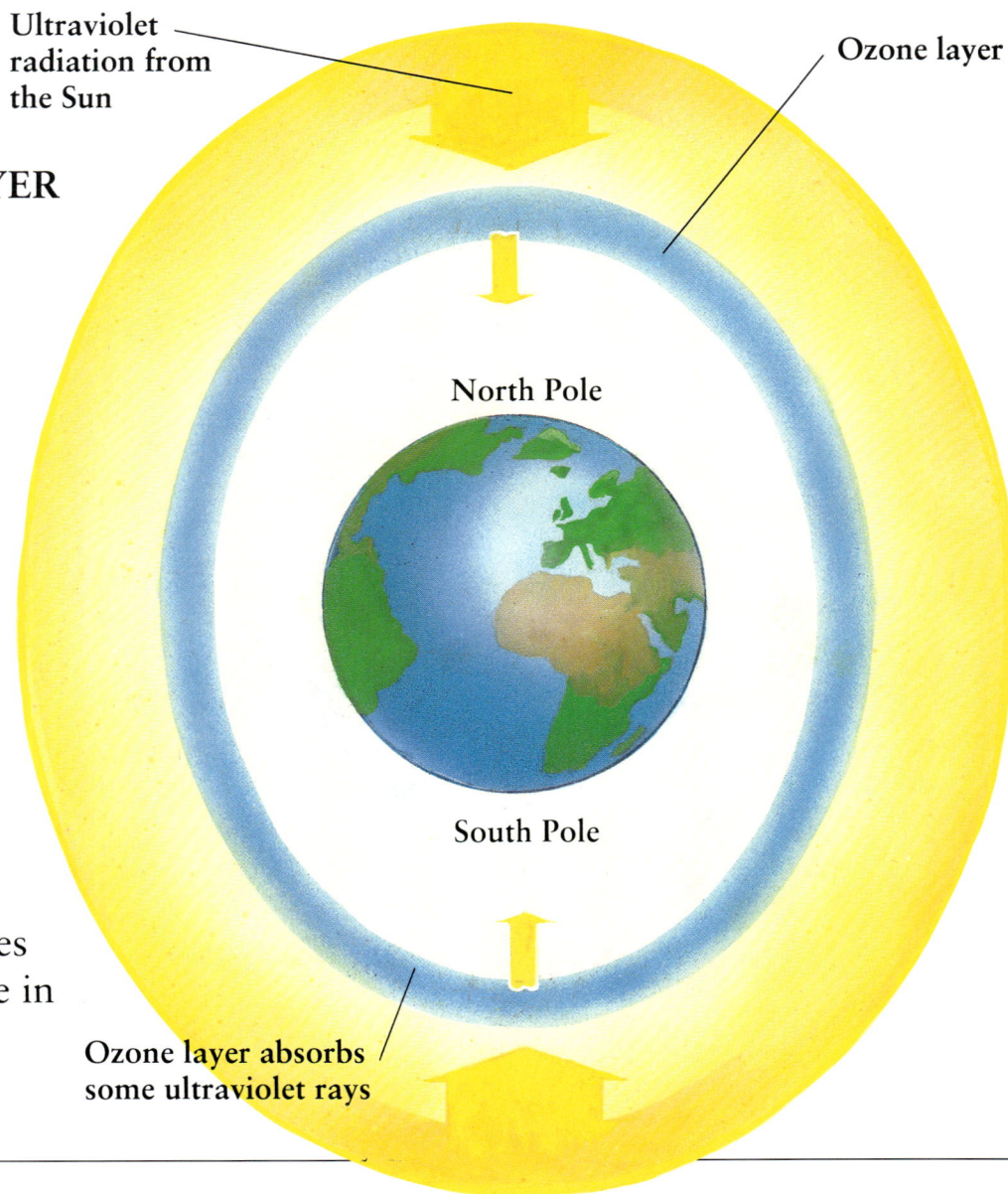

Ultraviolet radiation from the Sun

Ozone layer

North Pole

South Pole

Ozone layer absorbs some ultraviolet rays

Some aeroplanes fly in the stratosphere, where the ozone layer is. Scientists say that emissions from the engines destroy ozone. As air traffic increases, the likelihood of ozone damage grows.

LAYER

The "hole"

Scientists call the thinning of ozone over the Antarctic a "hole". It is not an actual hole, but when half or more of the ozone is destroyed, scientists talk about a "hole" in the ozone layer. The monitoring of the ozone layer can be difficult because ozone levels are affected by weather.

THE BRITISH ANTARCTIC SURVEY

Since the discovery of the "hole" over the Antarctic over ten years ago, the ozone layer has been closely monitored by scientists in Antarctica at the B.A.S. station (above). They know that there is 40% less ozone there now than there was 30 years ago.

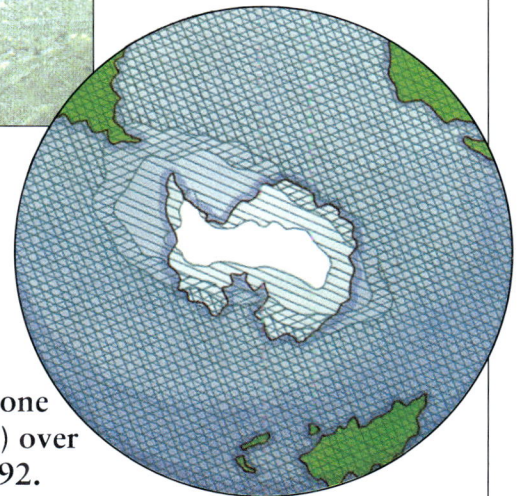

Picture of the ozone "hole" (in white) over Antarctica in 1992.

11

North Pole with thinning ozone band

South Pole with "hole" in ozone layer

Is there an Arctic "hole"?

There is as yet no "hole" over the Arctic, because weather conditions are different to the Antarctic; but there is depletion, or thinning of the ozone layer.

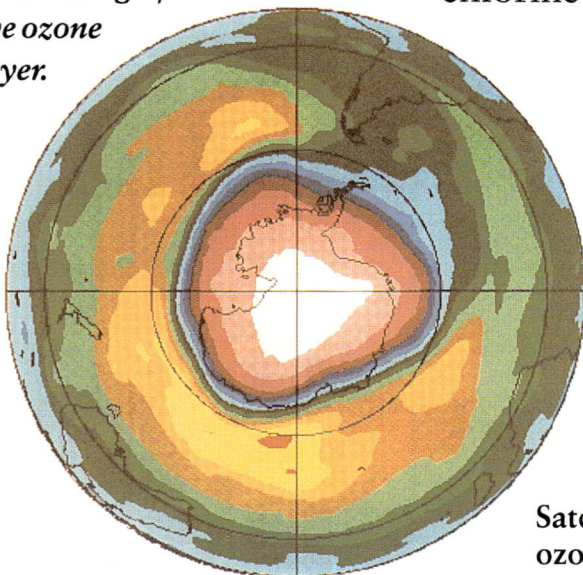

In 1985, the British Antarctic Survey announced the discovery of a "hole" in the ozone layer over the Antarctic which was particularly evident during the spring. Ten years later, in 1995 B.A.S. scientists reported that ozone was also starting to decline in the summer.

WHAT IS THE

HOW IS OZONE DESTROYED?

The "hole" in the ozone layer is caused mainly by CFCs (chlorofluorocarbons) – artificial chemicals in the atmosphere. CFCs are broken down by sunlight, causing them to release chlorine atoms. Each chlorine atom attacks ozone molecules. As many as 100,000 ozone molecules can be destroyed by one chlorine atom.

The Sun's damaging UV rays can reach the Earth's surface more easily through the depleted ozone layer.

Chlorine atoms remain unchanged and continue to destroy more ozone molecules.

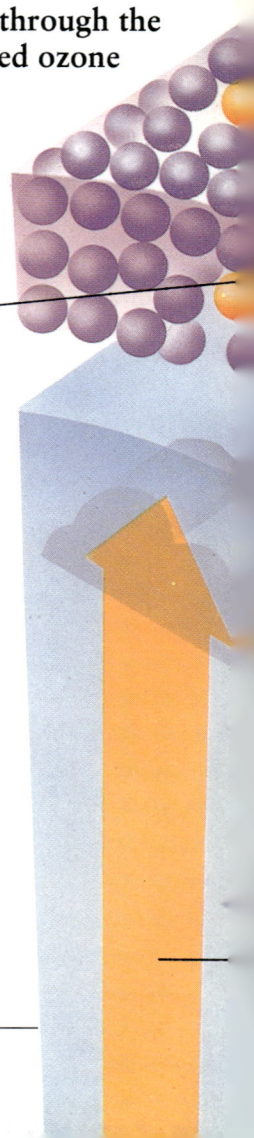

SATELLITE

This satellite photograph shows the ozone "hole" as a white area over the Antarctic. In 1994 the "hole" was reported to be 24 million km^2 – an area bigger than Europe.

Satellite picture of the ozone "hole" in 1994.

The Antarctic atmosphere has the ideal weather conditions for ozone destruction. In the cold, dark winter, chemical reactions occur in the clouds. These destroy ozone when sunlight returns in spring.

OZONE HOLE?

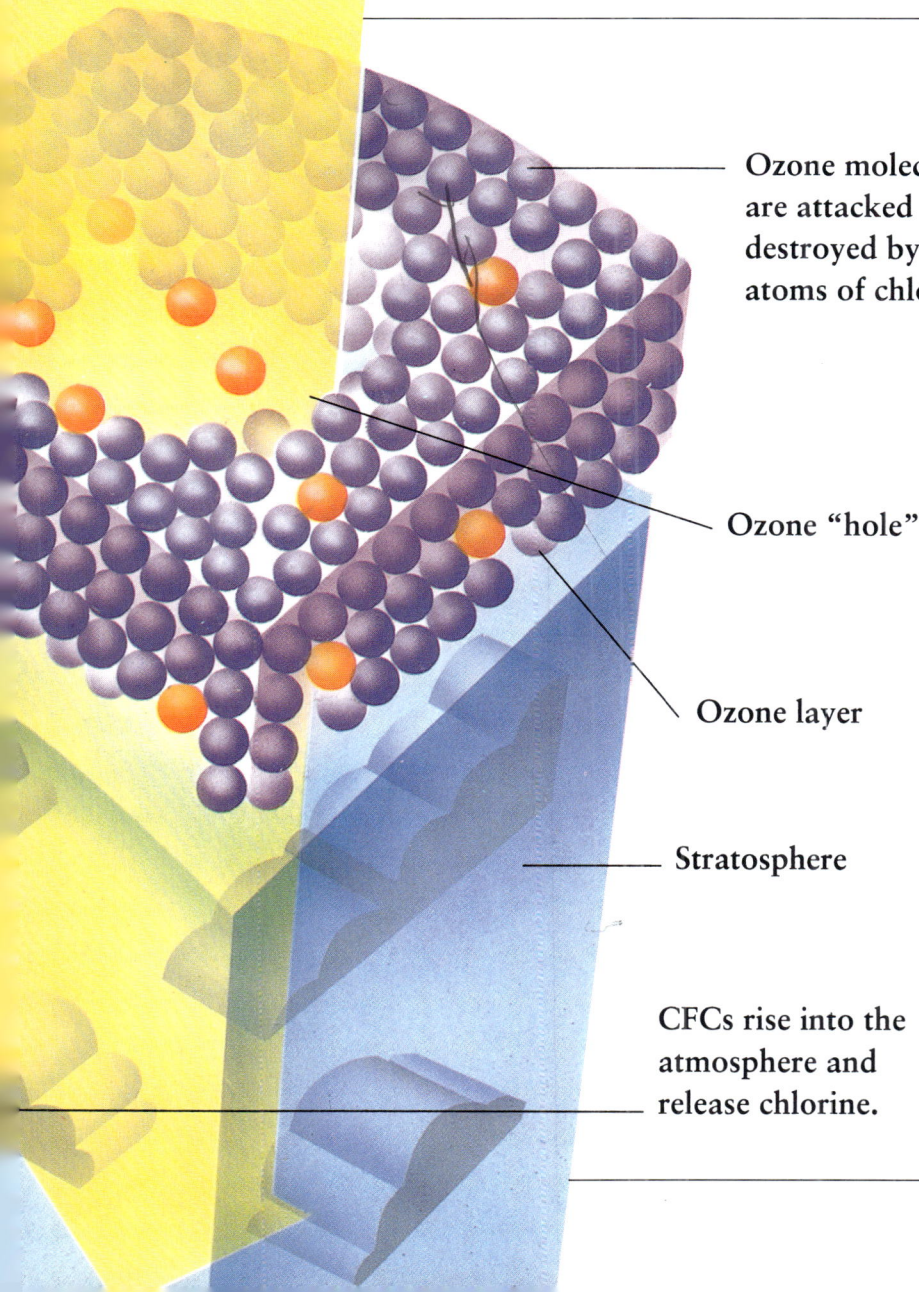

Ozone molecules are attacked and destroyed by atoms of chlorine.

Ozone "hole"

Ozone layer

Stratosphere

CFCs rise into the atmosphere and release chlorine.

Can we see it?
Ozone destruction is not visible to the human eye. Scientists need equipment like satellites to see it. If you were in the Antarctic, or sitting by an igloo in the Arctic, the sky would look normal.

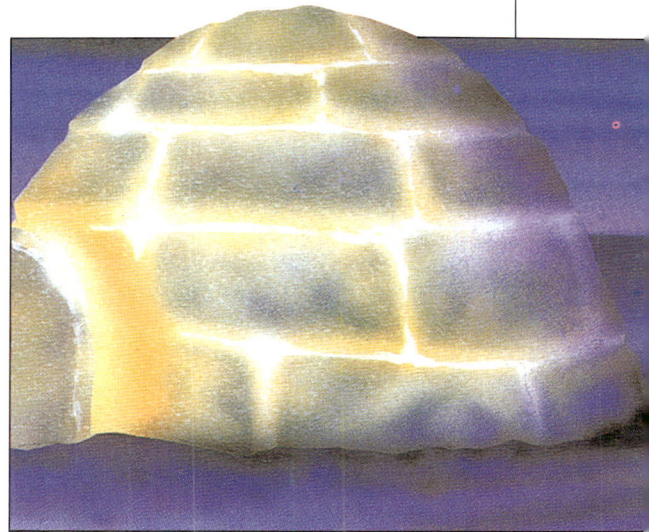

Smog

Ozone in the lower atmosphere can be dangerous. When sunlight mixes with gases found in exhaust fumes and factory emissions, ozone is formed. This makes up part of photochemical smog, which is harmful to people.

Ozone is a bluish-coloured gas and a form of oxygen. Most ozone is found in the ozone layer in the stratosphere. But ozone is also formed in the lower atmosphere, where it contributes to air pollution. Ozone levels vary at different times of the year, according to the strength of the Sun's rays reaching Earth.

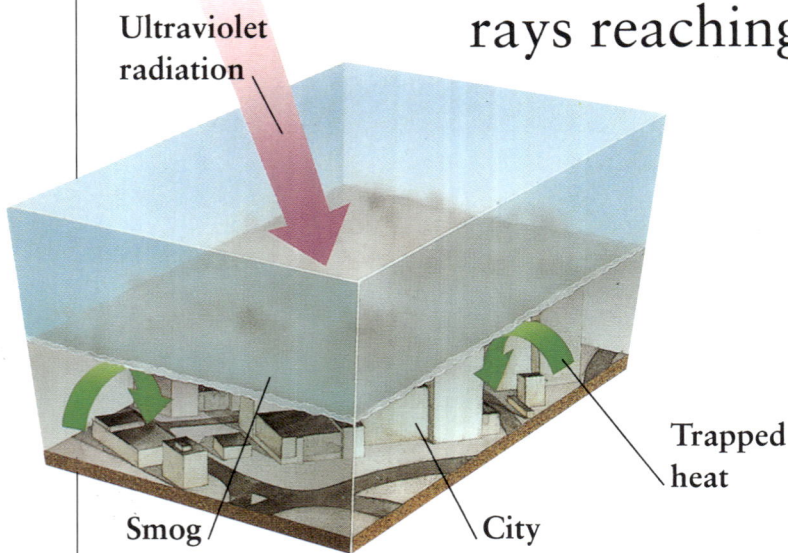

Ultraviolet radiation

Smog

City

Trapped heat

OZONE –

HEALTH HAZARD

Ground-level ozone is a different problem to the thinning of the ozone layer in the stratosphere. Photochemical smog (below left) is very common in big cities where there are lots of cars. The smog can cause breathing problems, eye irritation and damage to the lungs.

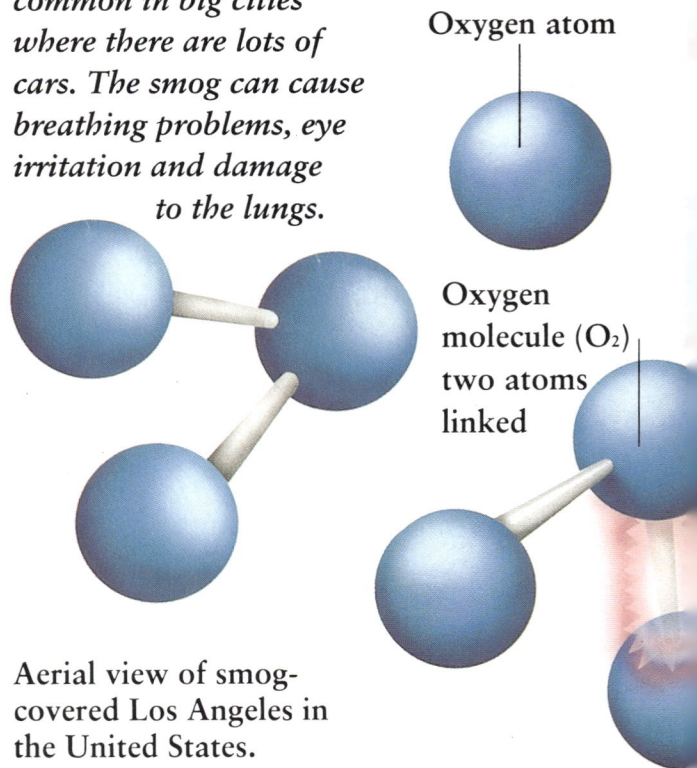

Oxygen atom

Oxygen molecule (O_2) two atoms linked

Aerial view of smog-covered Los Angeles in the United States.

Ozone makes up a small proportion of the greenhouse gases. In their natural quantities these gases are necessary for life. But as human-made pollution increases quantities of greenhouse gases, the Earth's climate may be affected.

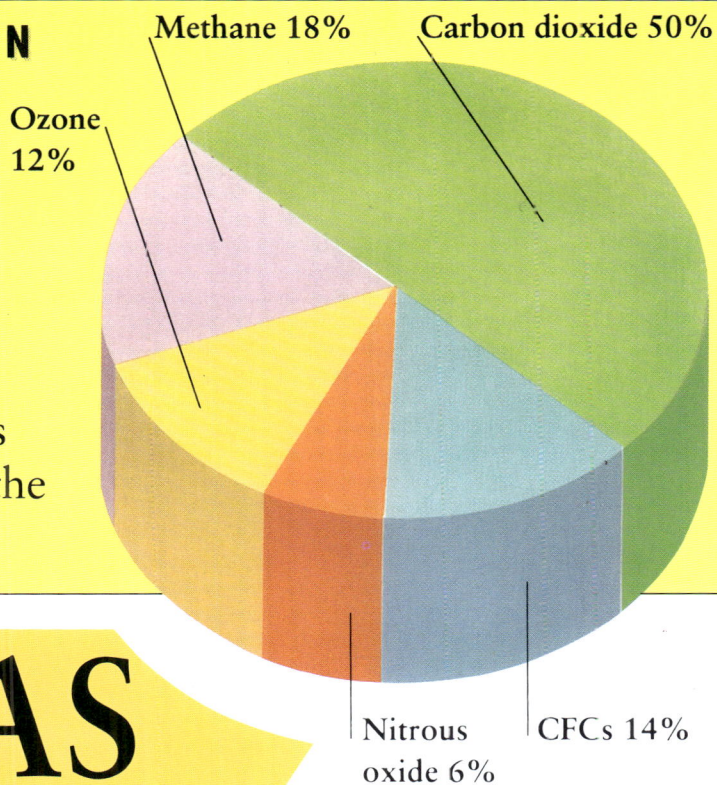

Methane 18%

Carbon dioxide 50%

Ozone 12%

Nitrous oxide 6%

CFCs 14%

IT'S A GAS

Sunlight

Ozone molecule (O_3) breaks apart

Ozone molecule (O_3)

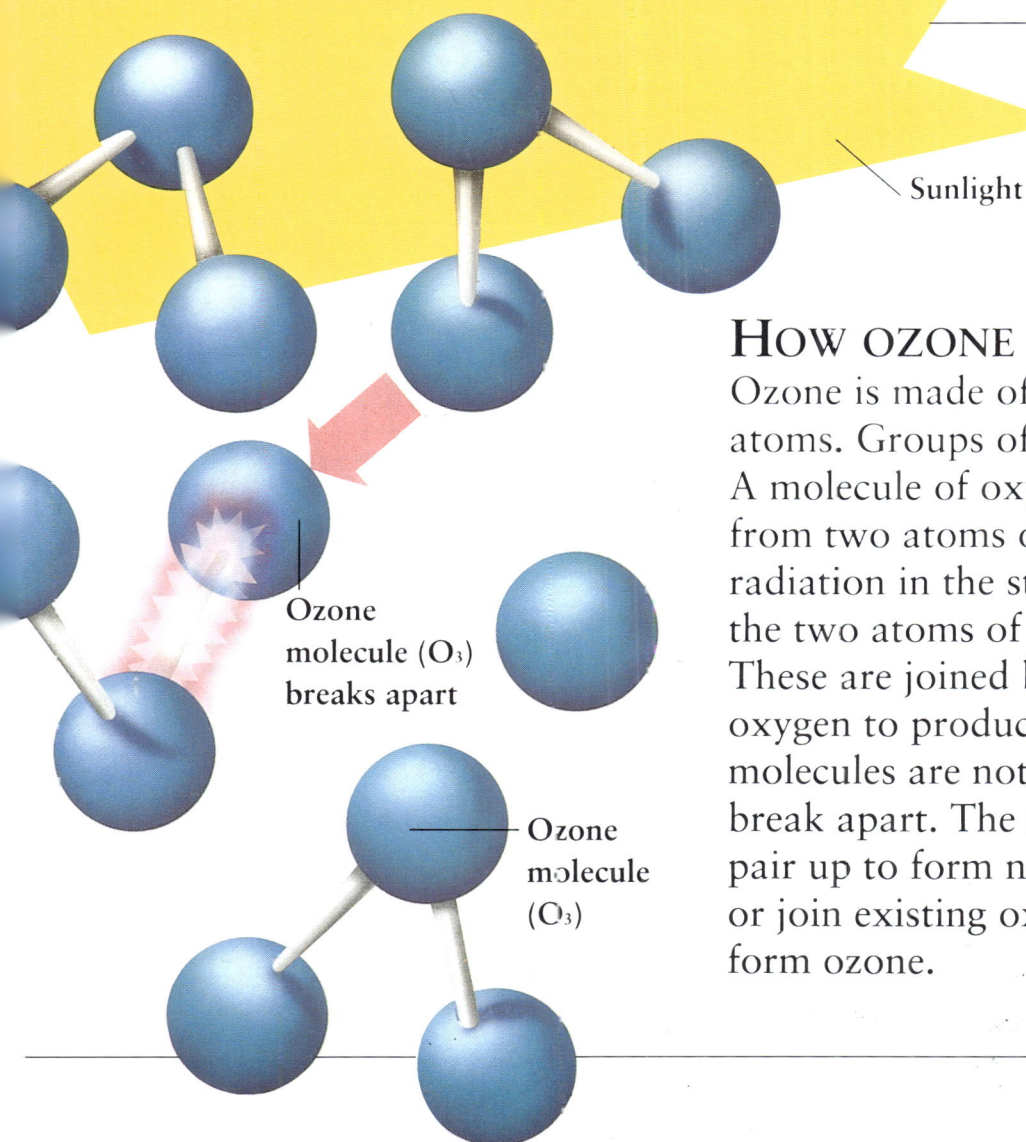

HOW OZONE IS FORMED

Ozone is made of tiny particles called atoms. Groups of atoms are molecules. A molecule of oxygen gas is made up from two atoms of oxygen. Ultraviolet radiation in the stratosphere separates the two atoms of the oxygen molecule. These are joined by a third atom of oxygen to produce ozone. These ozone molecules are not stable, and often break apart. The single atoms either pair up to form new oxygen molecules or join existing oxygen molecules to form ozone.

Skin Pigment

South-east Asian

European

Asian

African

Dark-skinned people tend to come from hotter areas of the world. The extra colouring, or pigment, in their skin protects them from the Sun. People with fairer skin generally come from northern regions where UV-B levels are lower.

The "hole" in the ozone layer allows dangerous UV-B rays to reach Earth. These rays have various harmful effects, some of which are not yet fully understood. The immediate effect on people is damage to health, particularly to skin.

SAVE OUR

KEEP OUT OF THE SUN

In moderation ultraviolet can be good for our bodies – it stimulates skin to produce vitamin D. But if we spend too much time in the Sun, we are more likely to suffer from sunburn, ageing of the skin and even skin cancer. Experts say that a 1% decrease of ozone in the ozone layer would lead to a 3% increase in skin cancer, and more cases of melanoma, which is the life-threatening form of skin cancer.

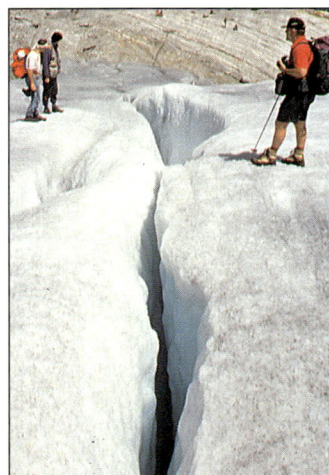

Sunscreens

It is best to use a sunblock when out in the Sun. It filters out harmful UV-B rays, and can protect skin from burning and from skin cancer. Climbers (left) use it; at high altitudes UV-B rays are more intense.

Some people want to look tanned and "healthy" all the time, and use artificial sunbeds. These beds use UV rays to darken skin, like the Sun, and can be just as dangerous.

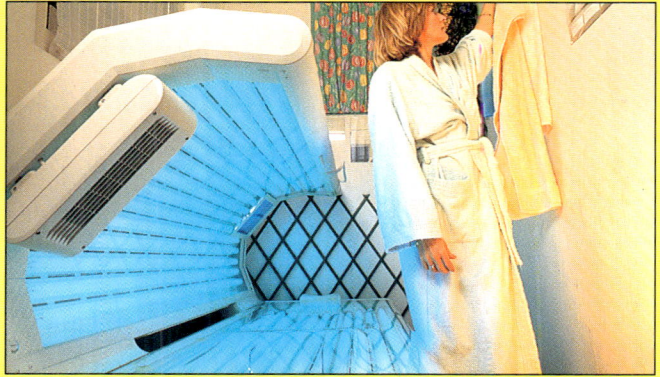

SKINS

OTHER POTENTIAL PROBLEMS

The increased amount of UV-B rays filtering through to Earth can cause eye disorders, such as cataracts. A cataract is the clouding of the lens of the eye which blurs vision (below). It is thought that UV-B rays may also weaken people's immune systems. The body's immune system helps us to fight off diseases. Weakening it would mean we are more vulnerable to illnesses.

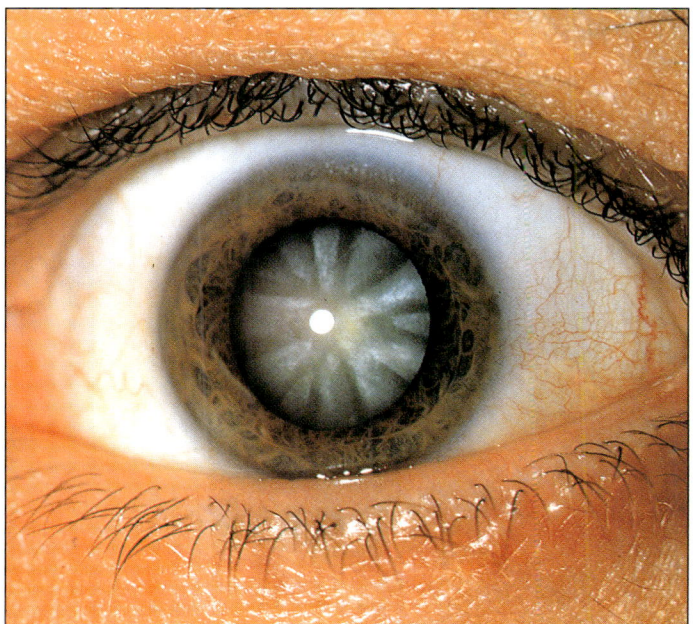

The marine food chain links plants and animals which rely on each other for food. Ultraviolet can penetrate 20 m into the sea, killing phytoplankton, a food source for many marine animals. If this link in the chain disappears many species may die out.

A BROKEN

Emperor penguin

Many animals feed on shrimp-like krill, a type of zooplankton, of which there are billions living in the Antarctic seas. Emperor penguins (above), blue whales, squid and fur seals all depend on krill for food, and would become extinct without it.

Krill

Whale

Fish

Phytoplankton

THE MARINE FOOD CHAIN

Phytoplankton are at the base of the marine food chain. Often called "the grass of the sea", they are microscopic plants, and are the main source of food for krill. Fish eat phytoplankton and,

On Closer Inspection – *Photosynthesis*

Green plants use energy from the Sun's rays for changing water and carbon dioxide into sugar for food. Too much UV light hinders plant growth and crop yield.

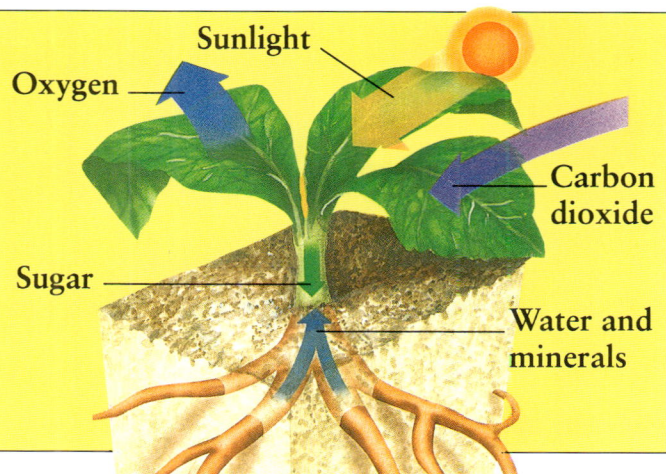

Sunlight
Oxygen
Carbon dioxide
Sugar
Water and minerals

CHAIN ?

Humans

Birds

FISH FAMINE

Phytoplankton are the main food source for fish, and fish are one of our main food sources (below). If fish lose their food supplies and die, the seas will not provide us with the food that helps us to survive. This will especially affect some developing countries where fish is a main ingredient in people's diet. It supplies essential protein in their diet.

n turn, birds eat the fish. The fish may then end up on someone's dinner able, thus completing the marine food chain. Blue whales, however, feed directly on krill.

Air cooling
Air-conditioning units have used CFCs as a cooling agent.

CFCs were invented in 1928 and became widely used in industry. In 1974 two American scientists first linked CFCs to ozone depletion. Now the use of CFCs is banned in industrialised countries. Despite this, 360,000 tonnes of CFCs were produced in 1995!

THE OLD

Fast food
CFCs have been used in the manufacture of foam packaging like fast-food containers.

Fridges
In the past CFCs in fridges were cooling fluids which moved through the fridge.

CFCs IN ACTION

On Earth, CFCs are inert (they never change and don't react with other chemicals). But as they move up into the stratosphere, they react with UV radiation. As they break up, they release chlorine atoms which can destroy ozone for many years.

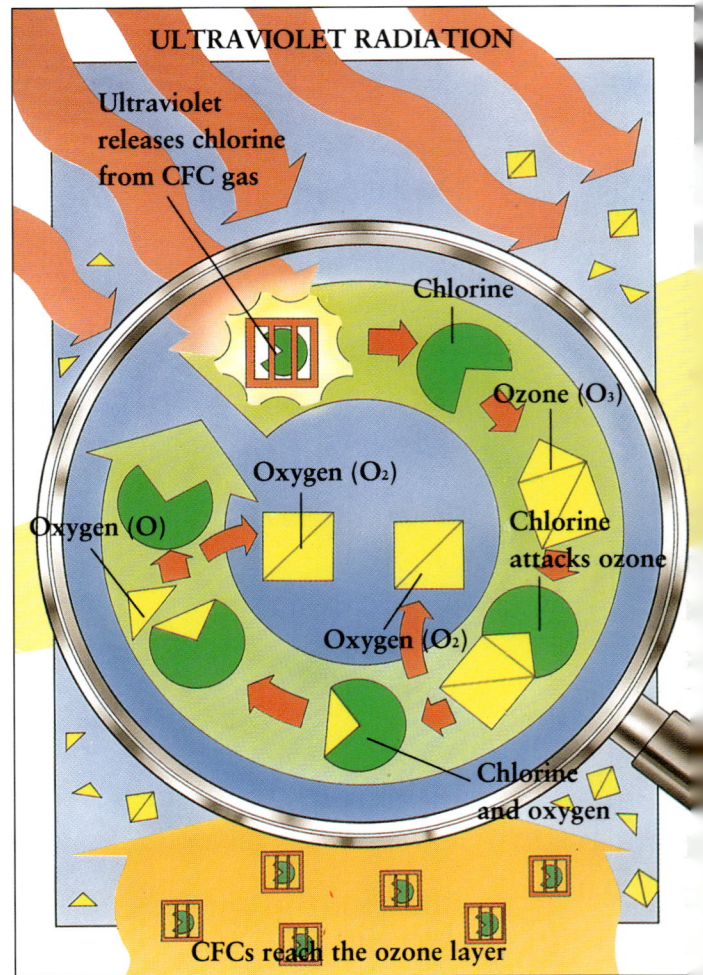

ULTRAVIOLET RADIATION

Ultraviolet releases chlorine from CFC gas

Chlorine

Ozone (O$_3$)

Oxygen (O$_2$)

Oxygen (O)

Chlorine attacks ozone

Oxygen (O$_2$)

Chlorine and oxygen

CFCs reach the ozone layer

ON CLOSER INSPECTION – *Aerosols*

CFCs were once always used as propellants in aerosols. 90% of aerosols are now free from CFCs and other ozone-damaging substances.

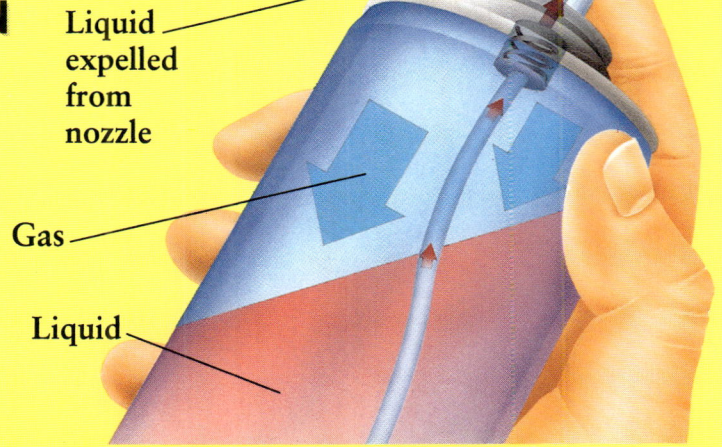

Liquid expelled from nozzle

Gas

Liquid

ENEMY

THE USES OF CFCs

CFCs are non-toxic, non-flammable and cannot be broken down easily – a quality which allows them to destroy ozone in the atmosphere for over 100 years. There are five main types of CFCs. This chart shows two of them (CFC-11 and CFC-12) and their uses. Until recently CFCs were used for cleaning, making solvents (liquids that dissolve substances) and for soldering metals. They have been useful for cleaning computer parts, because they do not damage the computer itself. Some CFCs are still produced for essential uses, and will be gradually phased out and replaced.

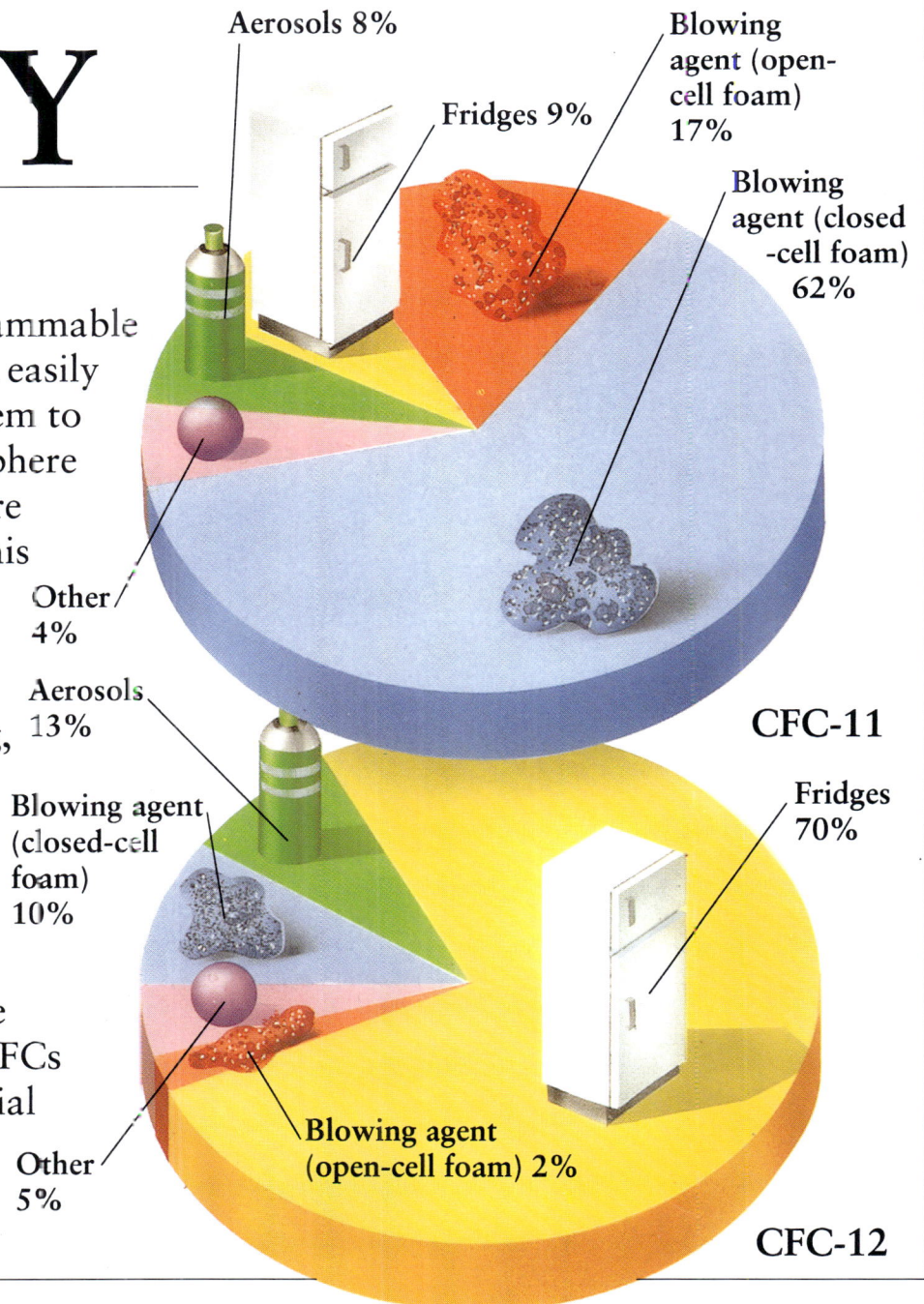

Aerosols 8%

Fridges 9%

Blowing agent (open-cell foam) 17%

Blowing agent (closed-cell foam) 62%

Other 4%

CFC-11

Aerosols 13%

Blowing agent (closed-cell foam) 10%

Fridges 70%

Other 5%

Blowing agent (open-cell foam) 2%

CFC-12

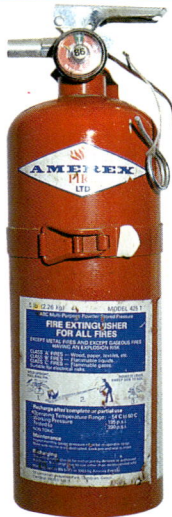

CFCs are not the only chemicals that destroy ozone. Many other ozone-depleting chemicals are used in industrial processes and everyday products. Bromine, which is contained in halons and methyl bromide, is an even more powerful ozone destroyer than chlorine from CFCs, and is thought to account for at least 25% of ozone loss over the Antarctic.

Where they come from

Halons are used in fire extinguishers (above). They put out fires, but don't harm people or machinery. Methyl bromide is used as a pesticide in farming (below).

MORE OZONE

INDUSTRY
Many chemicals which threaten the ozone layer come from factories and industry. Carbon tetrachloride is toxic and its use has been reduced or prohibited in many countries. It is still used in the production of essential CFCs, but in a safe way, so that it does not escape during production.

ON CLOSER INSPECTION
– *Natural causes*

Harmful chemical reactions can also occur naturally. When Mount Pinatubo (right) erupted in 1991, 22 million tonnes of sulphur dioxide were sent into the stratosphere. These sulphur particles are involved in chemical reactions which release chlorine – a powerful ozone destroyer.

- EATERS

TABLE OF OZONE ENEMIES

This table shows some of the main ozone destroyers – their uses, how long they stay in the atmosphere and their ozone-damaging potential. For many, alternatives have been found that are more ozone-friendly. For some, like methyl bromide, it is proving more difficult to find alternatives.

Taking CFC-11 as one (◆), units of ozone damage potential are in relation to this.

Substance	Lifetime in Atmosphere	Ozone Damage Potential	Uses
CFC-11	60 years	◆	Refrigeration and air conditioning
CFC-12	120 years	◆	Refrigeration and air conditioning
Halons	25-110 years	◆◆◆ to ◆◆◆◆◆◆◆◆◆◆	Fire fighting
Methyl Bromide	1.3 years	◀	Pest control
Carbon Tetrachloride	50 years	◆	Laboratory uses

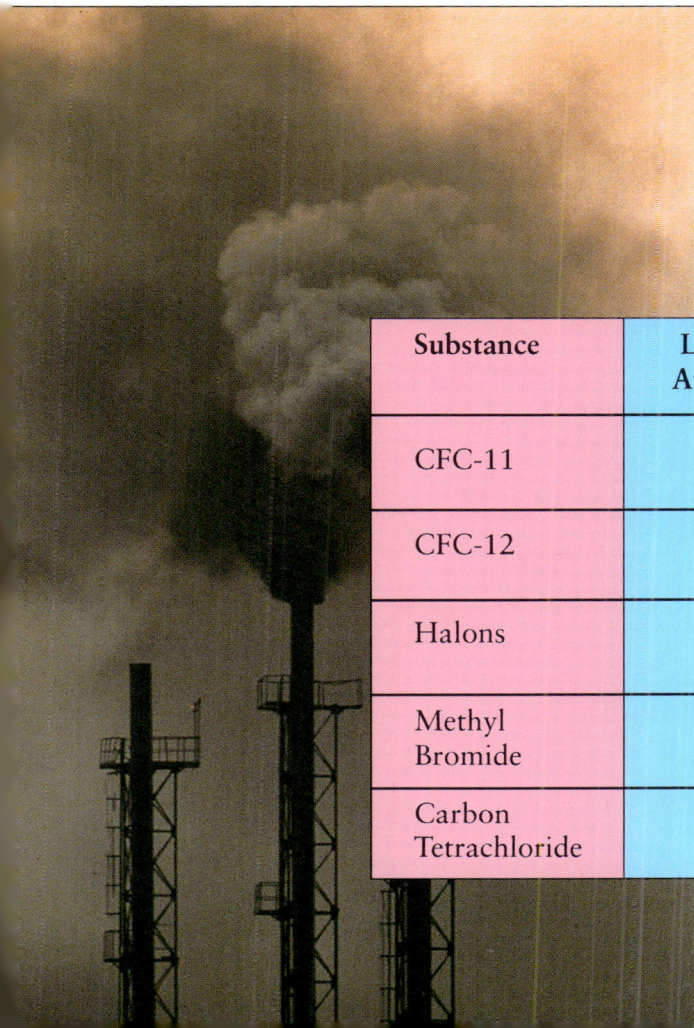

In 1957, the British Antarctic Survey started to measure ozone levels in the Antarctic. The damage to ozone is continuing; as recently as 1995, scientists at the Halley Research Station in Antarctica reported that the "hole" is still deepening.

HOW BIG IS

Antarctic story

In 1985, Joe Farman from the B.A.S. station discovered the "hole". Many research methods are used to study climate, including taking ice cores from the snow to show weather patterns from the past (above).

Targeting the problem

Since CFCs were identified as ozone destroyers alternatives have been developed. The label above indicates there are no CFCs in the product.

HOW EASY IS IT TO TACKLE?

Until recently, it was mainly industrialised countries who produced CFCs, and who were under the most pressure to ban them. Now developing nations are being urged to do the same, but CFC products are relatively new in these countries and are in increasing demand. Industrialised nations are giving money to developing countries to help them phase out CFCs and develop alternatives – a process which is very expensive.

Fridges (containing CFCs) are increasingly sought after in developing countries.

THE PROBLEM?

A GROWING PROBLEM

Ozone measured annually since 1979 has decreased by an average of 9% a year. In 1994, nearly all the ozone around the Antarctic between the altitudes of 14 and 18 km had disappeared. For ten days in that year, the ozone "hole" extended to the tip of South America. Evidence from satellites today indicates that the ozone layer is now thinning across the world.

Representation of satellite picture of 1994 ozone "hole" extending over the tip of South America.

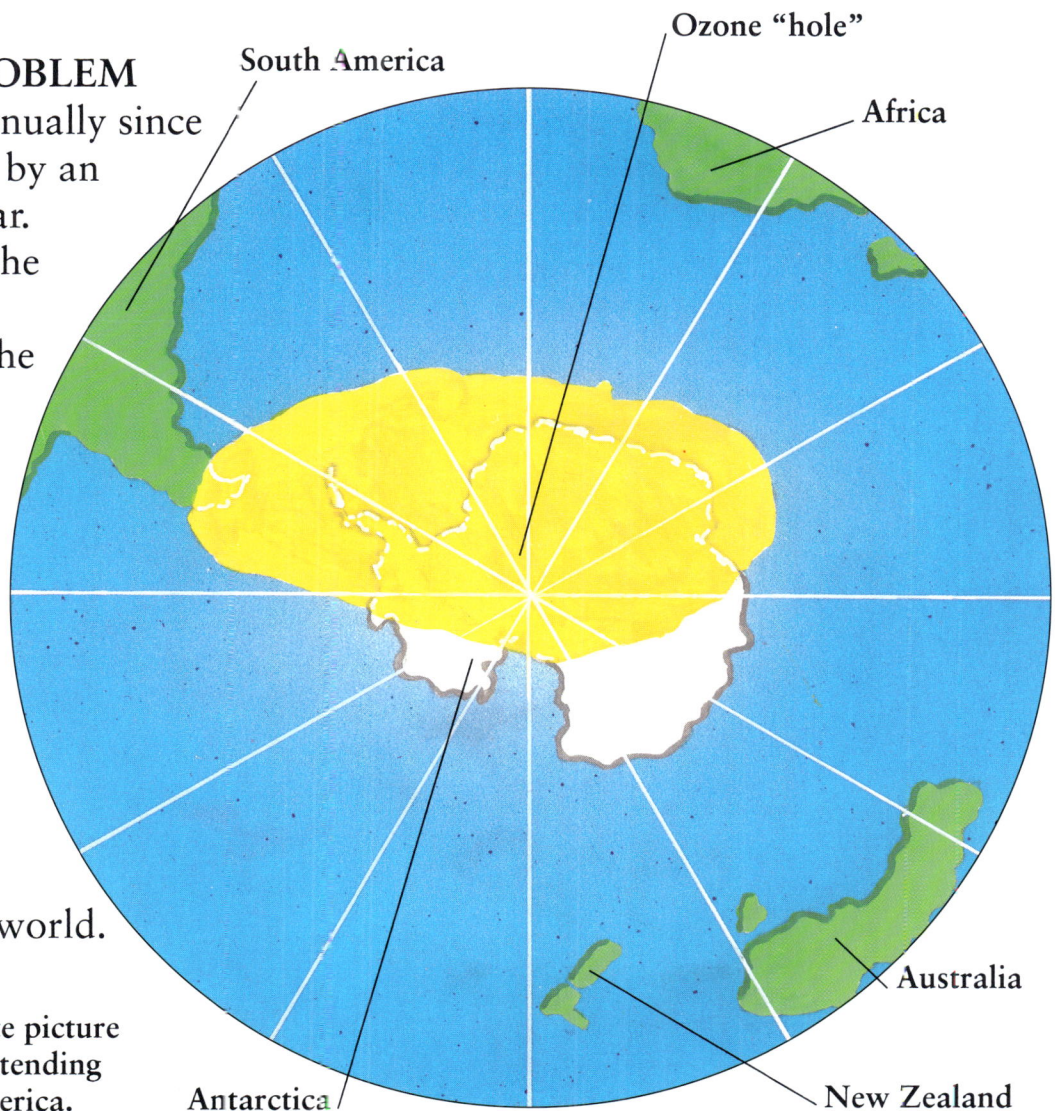

South America

Ozone "hole"

Africa

Australia

New Zealand

Antarctica

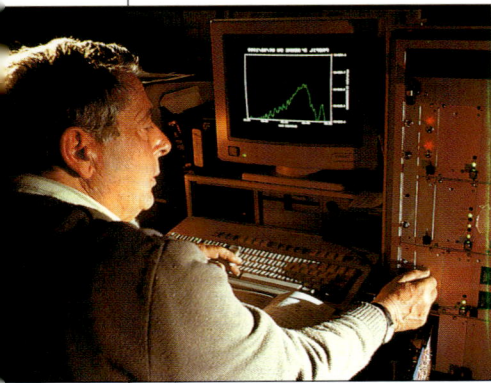

ow that scientists and politicians are aware of the threat that the thinning in the ozone layer presents, the most modern scientific equipment is being used to track the ozone "hole". The more that we understand about ozone and the ozone layer, the more we will be able to prevent further damage to the ozone layer, the atmosphere and our Earth.

Ozone profile
This French scientist at the Haute-Provence observatory is looking at a graphic display of the ozone profile. A laser is fired into the atmosphere at night to gather this data.

TRACKING

WEATHER BALLOONS

Weather balloons are used in the atmosphere for research. They carry measuring and other experimental equipment high up into the atmosphere. These balloons can fly up to 28 km above Earth. In 1990, scientists from France, Germany and the United States launched balloons like this (left) to monitor ozone. In 1992, similar balloons monitored ozone depletion for the European Arctic Stratospheric Ozone Experiment.

The launch of a weather balloon carrying observatory equipment.

Scientists looking at the Antarctic "hole" have also kept the Arctic under constant surveillance. Important chemical changes have been noticed in the stratosphere in northern regions. Ozone depletion (caused by pollution) stands at 6% to 9% around the Arctic, in the early months of the year.

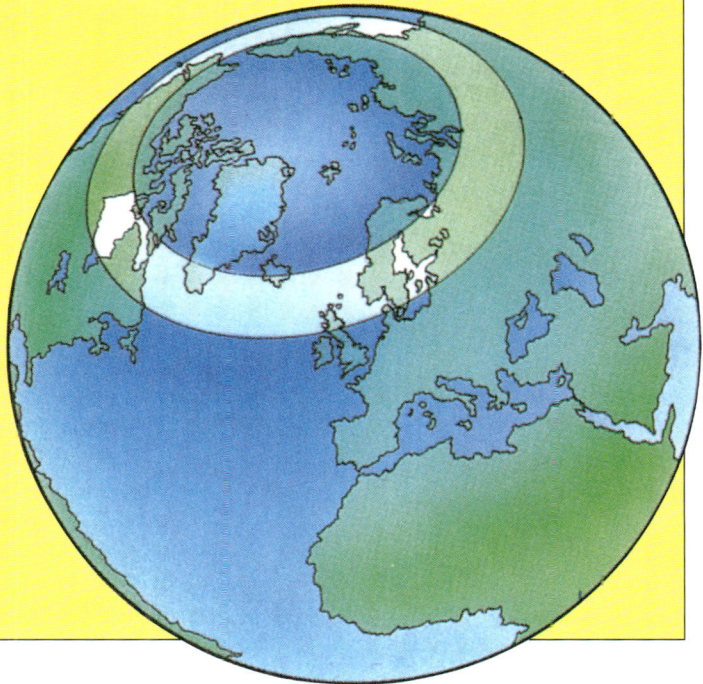

THE HOLE

SPECTROPHOTOMETER

The spectrophotometer (below) measures the radiation reaching Earth, which relates to how much ozone is in the atmosphere. This scientist is filling the instrument with liquid nitrogen for an experiment in the Arctic.

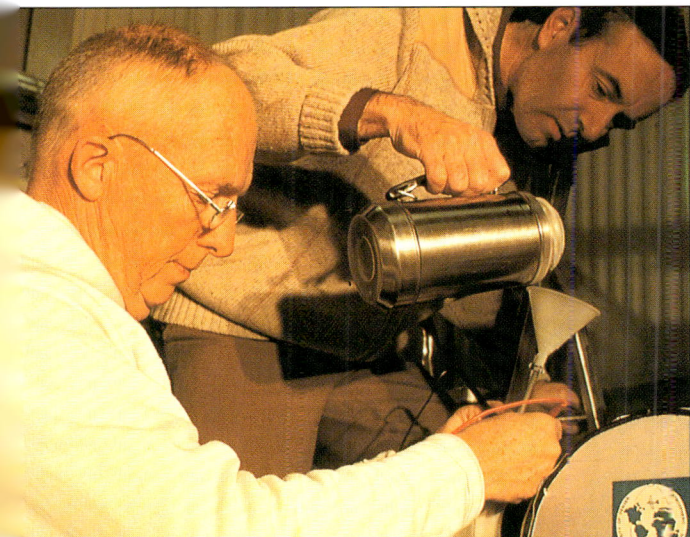

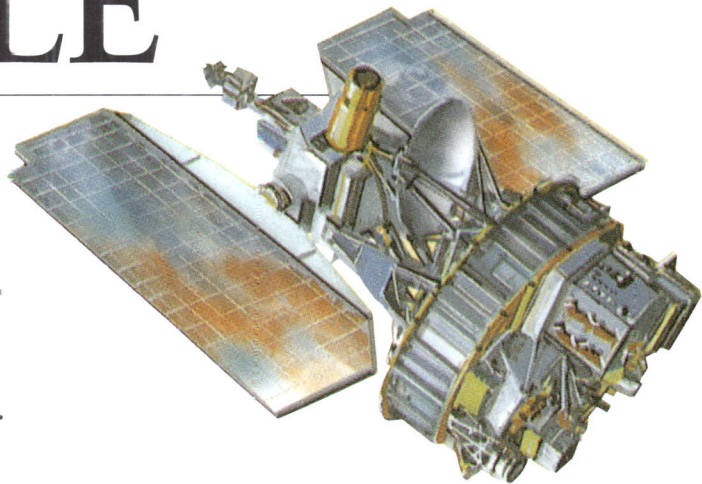

SATELLITES

The Nimbus 7 weather satellite (above) was launched in 1978 by NASA (National Aeronautics Space Administration) based in the U.S.A. It had eight different instruments on board to monitor weather patterns. This and other satellites regularly orbit the Earth, taking pictures with cameras, and transmitting information to scientists back on Earth.

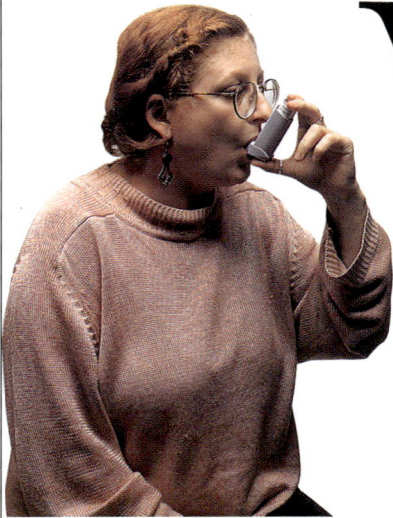

Exceptions to the rule

There are some uses for CFCs that are essential. For example, some inhalers, which are used by asthma sufferers (above), use CFCs as propellants.

MONTREAL PROTOCOL

By December 1995, 150 countries had signed the Montreal Protocol. In 1987, they first met to discuss substances that deplete the ozone layer. Since then they have met another six times. They set 1995 as the phase-out date for CFCs and agreed dates for phasing out other ozone-depleting chemicals. Developing countries were given an extra ten years to phase out ozone-depleting chemicals.

e have already added over 320 million tonnes of CFCs to the atmosphere, and some will stay there well into the next century. This means that damage to the ozone layer will continue until at least the year 2000. In some developing countries, CFC use is still on the increase. It is estimated that China used 80% more CFCs in 1995 than in 1993.

MENDING

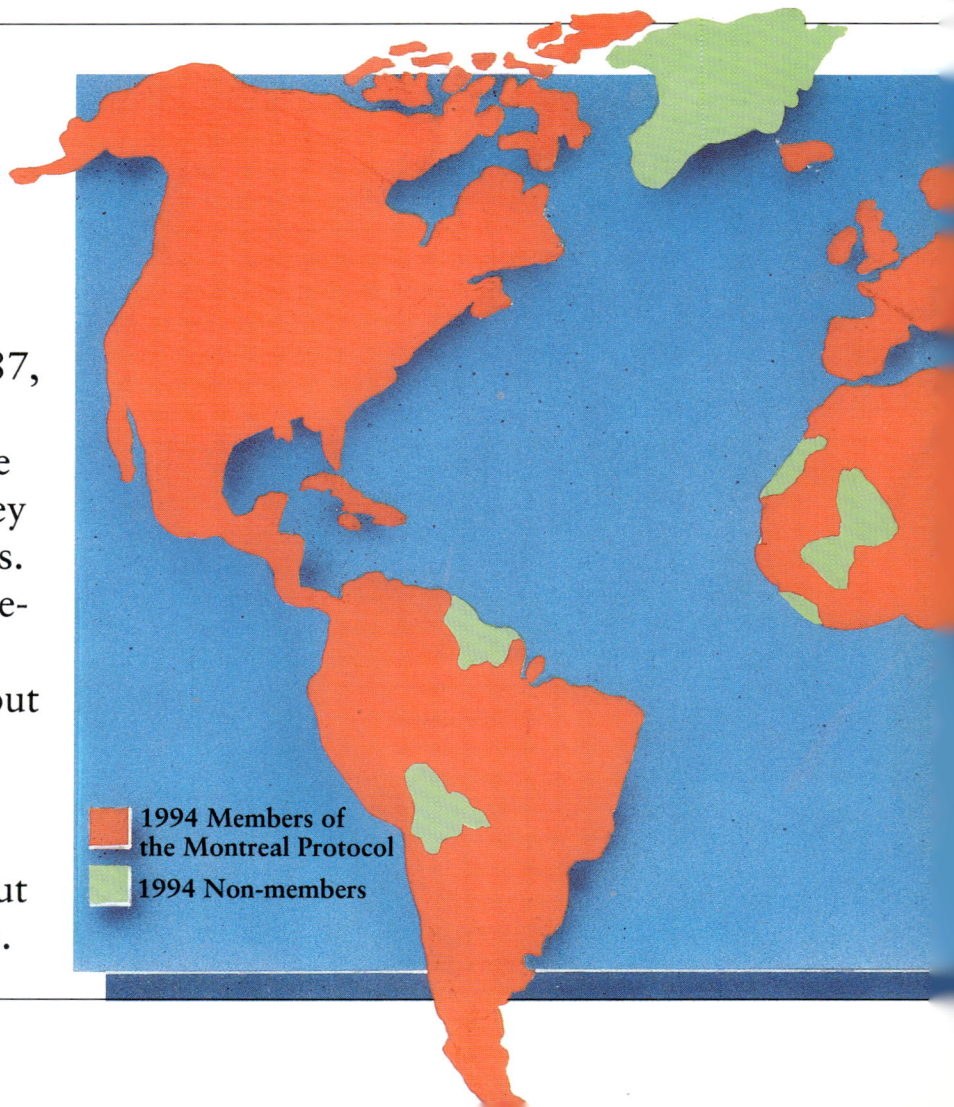

■ 1994 Members of the Montreal Protocol

■ 1994 Non-members

ON CLOSER INSPECTION
– Future ideas

Scientists are working on ideas to stop ozone destruction, including one which uses laser beams to destroy CFCs. Other ideas include weakening CFCs by pumping gases into the atmosphere, and taking ozone at ground level, where it is harmful, and adding it to the ozone layer.

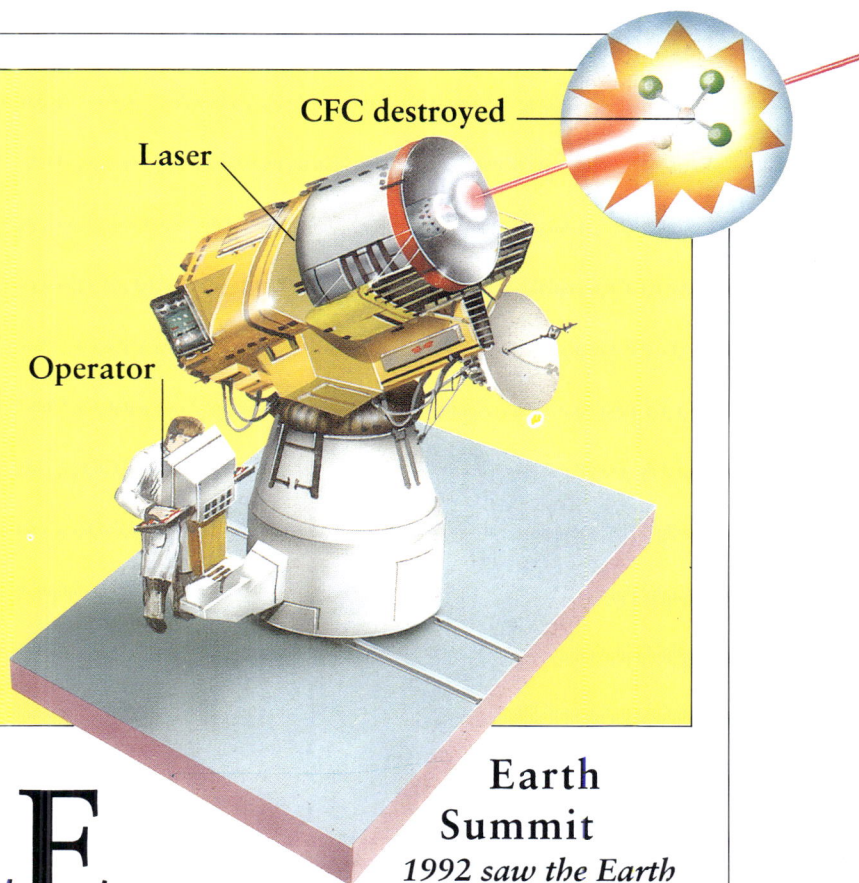

CFC destroyed

Laser

Operator

THE HOLE

Earth Summit

1992 saw the Earth Summit in Brazil, where leaders from around the world discussed problems of the environment for the first time.

The opening of the 1992 Earth Summit in Brazil.

GET INFORMED

Read all you can about the ozone layer. Get your information from environmental groups such as Friends of the Earth and Greenpeace, as well as the Department of the Environment and newspapers and magazines.

FRIENDS *of the*
earth

GREENPEACE

WHAT CAN YOU DO?

RECYCLE!

It is now possible to recycle fridges and fire extinguishers which use ozone-depleting chemicals. CFCs are removed from old fridges, cleaned, recycled and put into new fridges. Encourage your family to recycle their old fridges and freezers instead of throwing them away

TROUBLE-FREE SPRAYS

Use ozone-friendly, pump-action spray cans which don't need a propellant gas of any kind; and check that your friends and family don't use ozone-eating sprays either.

TO SUM IT ALL UP...

It's not just up to huge chemical companies and governments to save the ozone layer; above are some of the things that we can all do to make a difference. Most of all, remember that although we have gone some way towards saving the ozone layer, there is still a lot of work to be done!

Atmosphere The mixture of gases that surrounds the Earth. It provides some of the conditions needed for life.

Atom The smallest particle of an element. Atoms combine to form molecules.

Cataract A cloudiness of the lens in the eye which results in blurred vision.

Chlorofluorocarbons (CFCs) Human-made chemicals that destroy ozone. They were previously used for a variety of applications: fridges, aerosols and foam packaging.

Electromagnetic spectrum The whole range of types of radiation, all of which travel at the speed of light: 300,000 km per second.

Greenhouse gases Gases which allow the Sun's energy to pass through the atmosphere to warm the Earth's surface, and then trap the heat as it is reflected back into the atmosphere.

Immune system A system within the body which helps you to resist disease.

Melanoma A cancerous skin tumour with a dark colour.

Molecule A group of atoms. The most basic part of a chemical compound or element.

Ozone A form of oxygen with three atoms, a bluish gas with a bitter smell.

Ozone layer A band of ozone in the atmosphere.

Photochemical smog A form of air pollution caused by the action of ultraviolet radiation on exhaust gases from motor vehicles and factories.

Photosynthesis The method that plants use for converting the Sun's energy into sugars so that they can grow.

Phytoplankton Tiny plants living in the seas.

GLOSSARY

Pigment A coloured substance. The chlorophyll pigment produces the green colour in plants. Melanin pigment produces the brown colour in our skin.

Propellant A substance used to force a liquid out of an aerosol as a spray.

Radiation The transferring of energy in straight lines.

Solar system The nine planets that circle the Sun make up the solar system, as well as other objects, like moons, asteroids, comets, dust and rocks.

Ultraviolet High-energy, short-wavelength light. This radiation, which is produced by the Sun, can cause suntans and damage to skin.

INDEX

Photo credits